I0392532

Computation Notebook
100 Pages Graph Paper

By, The Notebook Gurus

CreateSpace Independent Publishing Platform

For the complete selection of notebooks, visit
www.thenotebookgurus.com

Subject: ______________________

Project: ______________________

Volume: ______________________

Name: ______________________

Address: ______________________

City: ______________________

State: ______________________

Country: ______________________

Phone: ______________________

Email: ______________________

Description: ______________________

____

____

Page	Topic	Date
1		
2		
3		
4		
5		
6		
7		
8		
9		
10		
11		
12		
13		
14		
15		
16		
17		
18		
19		
20		
21		
22		
23		
24		
25		

Page	Topic	Date
26		
27		
28		
29		
30		
31		
32		
33		
34		
35		
36		
37		
38		
39		
40		
41		
42		
43		
44		
45		
46		
47		
48		
49		
50		

Page	Topic	Date
51		
52		
53		
54		
55		
56		
57		
58		
59		
60		
61		
62		
63		
64		
65		
66		
67		
68		
69		
70		
71		
72		
73		
74		
75		

Page	Topic	Date
76		
77		
78		
79		
80		
81		
82		
83		
84		
85		
86		
87		
88		
89		
90		
91		
92		
93		
94		
95		
96		
97		
98		
99		
100		

www.ingramcontent.com/pod-product-compliance
Lightning Source LLC
Chambersburg PA
CBHW080707190526
45169CB00006B/2275
* 9 7 8 1 5 3 7 4 6 3 6 8 1 *